The Chimpanzee

The Living Link between 'Man' and 'Beast'

JANE GOODALL

THE THIRD
EDINBURGH MEDAL
ADDRESS

© Jane Goodall 1992
Reprinted 1994

Set in Linotron Janson by
Nene Phototypesetters Ltd
and printed in Great Britain by
Alexander Ritchie & Son, Edinburgh and
bound by Hunter and Foulis, Edinburgh

ISBN 0 7486 0354 9

Frontispiece: Jane Goodall with the chimp, Whiskey.

The Lecture

I need hardly say that it is a tremendous honour for me to be here tonight to be presented with this medal. I want to assure you that this prize will be used to further the studying of the chimpanzees of Gombe, chimpanzees in other parts of Africa, and chimpanzees in captivity. The prize money will also be used to help conservation efforts so that we may continue to study chimpanzees in the future. If we don't protect them, there is a very real danger, as will become apparent, that scientific studies of wild chimpanzees in Africa will come to an end.

I have links with Edinburgh. It was Professor Aubrey Manning and Dr Anthony Collins who first brought me here, to speak for the Zoology Department. (Tony Collins is at Gombe now, coping with the day-to-day running of the research station as well as carrying on with his baboon research.) And, three years ago, we launched the Jane Goodall Institute (the JGI) in this very room. So Edinburgh is a place of happy associations for me.

When I was a child, growing up in Bournemouth, I had two dreams. From the start I wanted to live among and study animals. And then, as I grew older and read more and more books about animals, I knew I wanted to go to Africa. And I wanted to write books about animals, to share what I learned with as many people as I could. I have been very fortunate: how many people can stand up at my age and look back over their lives and know that they have indeed realised their childhood dreams?

I have been incredibly fortunate also in being able to spend thirty years living among, learning about – indeed, learning *from* – our closest living relatives. Chimpanzees are closer to us genetically than any other living being. They share with us over 98 per cent of our DNA, our genetic material. Chimpanzees are closer to humans in these ways than they are even to gorillas. When I was planning this talk I spent some time wondering how best to share something of the rich experiences of the past thirty years. When I lie in bed, thinking, the memories just tumble over each other. I finally decided that it would be best to take you with me on a little journey to Gombe, to describe something of the history of a community of beings who cannot write it for themselves, and to introduce to you some of the wonderful personalities that I have had the · privilege of knowing.

The Gombe National Park, where our research centre is located, is a very tiny area, the smallest of Tanzania's national parks. Tanzania has had a wonderful conservation record, with twenty-five per cent of her land surface under conservation status of one sort or another. Gombe is a little jewel of a national park, but it is only thirty square miles. It rises up from the eastern shore of Lake Tanganyika, opposite Zaire, extends inland only to the peaks of the rift escarpment, and runs for only about ten miles along the shore of the Lake. There are thickly forested valleys with sparsely forested or grass covered ridges and peaks in between.

When I first got to Gombe in 1960 the worst problem that I had to overcome was the fact that the chimps were so shy that they ran off even if I was on the far side of a valley. But gradually, because I wore the same coloured clothes and did not try to get too close to them too quickly, they came to realise that I wasn't so terrifying after all. And even in those early days when the chimps were still so shy, I was able to start piecing together something of their daily behaviour patterns. I spent much time sitting on one of the high peaks, watching through my binoculars. When I look back, I am not at all surprised that it took so long to understand their social structure – because it is incredibly complex. The typical social group, known as a community, comprises about fifty individuals (the average number over the years at Gombe). Of these, about eight to ten are fully adult males, rather more are adult females, and the rest are adolescents, juveniles and infants. But community members do not travel around in a troop, stable or semi-stable, as do baboons, gorillas and most monkeys. Rather they travel in temporary associations, perhaps a group of two females and their families. These two females will remain together while they are feeding, then they may spend the rest of the day wandering about peacefully while their young ones play. Then they may separate for the night. Next day, one female may stay on her own with her children, the other may join a lone male or a group of males or another female and her family or a mixed group. There is continual coming and going as small groups form and separate and re-form again. Sometimes there are large excited gatherings when many members of the community join together for a while – as when there is a delectable food available in one part of the range, or when there is a highly sexually popular female (and there is great variation between females as to how many males they attract when they are in oestrus).

After several months I began to identify different chimpanzees as individuals. (In those days, thirty years ago, it wasn't terribly fashionable to talk about animal personality in scientific circles, but fortunately I did not know that!) The first chimpanzee I came to know really well was David Greybeard. He had an unusually trusting disposition, and he allowed me closer than the others. His companions watched his acceptance of me with wide eyes, and gradually came to realise that I wasn't so frightening after all. It really was David, this wonderful chimpanzee, who opened for me the door into the magic world of the wild chimpanzees. One day, when I got down from the mountains, my Tanzanian cook told me that a chimpanzee had come to feed on the ripe fruit of an oil nut palm that grew there. This happened several days in succession, and one day this chimp took some bananas that were lying outside my tent. I determined to wait in camp the next day. The chimpanzee duly arrived; it was, as I had guessed, none other than David Greybeard. I had put out some bananas, and he came very close to take them. I shall never forget the first time that he actually took a banana from my hand. A fully grown adult male, a wild chimpanzee who had spent the earlier part of his life fearful of humans. And now he trusted me to the extent that he would take an offering from my hand. It laid a great responsibility on me, all those years ago, because once this trust had been established, I knew I should never forgive myself if I ever allowed it to be broken.

It was David Greybeard who taught me that chimpanzees in the wild not only use objects as tools but actually *make* tools. It was incredibly exciting when I first observed this, when I watched as David Greybeard pushed a grass stem into a termite mound (into a passage that he had scratched open), carefully withdrew it and picked off the termites that clung to it. Presently he picked a leafy twig: before this could be used David had to strip the leaves. Thus, by modifying the twig, he was showing crude tool making behaviour. When I was young we were taught that one important distinction between 'man' and 'beast' was that only we were capable of *making* tools. When I sent a cable about David's termite fishing exploits to my mentor, Dr Louis Leakey, he made his now famous remark

'Ah, now we must redefine *tool*, redefine *man* – or accept chimpanzees as humans!'

It was David, too, who taught me that chimpanzees not only eat fruits, leaves, blossoms, stems and insects, but that they are also efficient hunters and eat meat.

Gradually David introduced me to his close companions. Goliath and David were almost always together. Possibly they were siblings, although they did not look alike. David was often with the old female Flo, known to anybody who has seen any of the National Geographic Specials, with her bulbous nose and ragged ears. In those days her daughter Fifi was about five years old. David and Goliath, Flo and Fifi, comprise a typical small mixed group.

The chimpanzees of a community communicate with each other across distance by means of a loud *pant-hoot* call. (Demonstration of the call.) The chimpanzees at the Edinburgh Zoo will demonstrate this call for you. Each chimpanzee has his or her own distinctive voice. If they hear a pant-hoot from the other side of a valley they know exactly who is calling. They must then make decisions: to reply or to remain silent, to join the caller(s), remain where they are, or move off in the opposite direction. It is due to the fluid nature of chimpanzee society that its members are continually forced to make decisions of this sort: to travel alone or with others; to go east and eat figs, or west on a hunting party; and so on.

Because chimpanzees are continually separating and reuniting again they have developed a whole series of postures and gestures that we can loosely describe as greeting behaviour. Most fascinating is the sometimes uncannily close similarities between the greeting behaviours of the chimpanzees and those shown by many different human cultures around the world. These include kissing, holding hands, patting one another on the back and embracing. Not infrequently there is some aggression when two groups of chimpanzees meet together, particularly when adult males are present in both groups. In this context, subordinate individuals may become fearful. Often they then seek contact with another chimpanzee – usually but not always a more dominant individual – and this provides reassurance and serves to calm them down.

Chimpanzee society is very clearly male dominated. Males are larger than females, they are more overtly aggressive, and they fight more often. These fights can look extremely fierce and the victim screams loudly. But it is rare for a fight between community members to last longer than quarter of a minute, and it is even more unusual for such a fight to result in serious injury. Many of these fights break out very suddenly, often for what appears a trivial reason. Afterwards the victim, even though clearly fearful of the aggressor, will almost always approach and adopt a submissive posture. In response the aggressor will usually reach out and bestow what we call a reassurance gesture – he will touch, pat, kiss or embrace the supplicator. Then the subordinate, quite visibly, relaxes. Social harmony has been restored. It is this sequence – aggression, followed by submission, followed by reassurance – that enables the members of the society to maintain, for the most part, such friendly relations with each other. And, of course, chimpanzees, like all social mammals, solve most of their disputes by means of threatening postures and gestures. A typical threatening behaviour, in a feeding situation, is the bipedal swagger. When an adult male swaggers thus at a subordinate his message is very easy to understand: 'You keep away from this food or else. . . .' Chimpanzees, particularly adult males, are ordered in a dominance hierarchy, each one knowing his place relative to that of every other. This means that threatening behaviour serves to solve disputes and eliminates, for much of the time, the need for actual fighting. This is adaptive, since during a fight either or both of the participants could be wounded.

The most impressive and the most effective threat behaviour is the *charging display*. Although females sometimes display, especially high ranking, assertive females, it is typically a male performance. During such a display the individual charges flat out across the ground, slapping his hands, and stamping his feet. He may leap up and shake and sway the vegetation. His hair bristles, his lips are bunched in a ferocious scowl. He may pick up and hurl a rock, he may drag a large branch or palm frond behind him. He is, in fact, making himself look larger and more dangerous than he may actually be and in this way he can very often intimidate a rival – even a rival who is larger and heavier than himself – without having to risk a fight. We have found, over the thirty years of study, that the young males who display the most frequently, the most impressively, and with the most imagination, are the most likely to rise quickly to a high position in the male dominance hierarchy.

During the thirty years there have been a succession of males who rose, for a while, to the top ranking, or alpha, position of their community. When I first became familiar with the different individuals, and began to understand something of their complex social structure, the alpha male was David Greybeard's close companion, Goliath. He maintained this top position (and the male chimpanzee does have to work hard to stay on top) by means of his very fast and dramatic charging display, and because he had a good deal of what I can only describe as 'courage'. He was eventually overthrown by a chimpanzee of similar age, Mike. When I first knew Mike he ranked low in the male hierarchy. At that time there were fourteen males in the community – more than at any subsequent period in Gombe's history – and eleven of them were higher ranking than Mike. Mike, however, had a very strong dominance drive, was very strongly motivated to improve his position, and he was endowed with unusual intelligence. Thus in his charging displays Mike made use of empty paraffin cans that he found lying around in my camp. He learned to keep two or even three of these cans ahead of him, hitting and kicking them, as he charged towards his rivals; not surprisingly, they usually got out of the way of this noisy performance. In this way, Mike rose to the top ranking position in just four months, and we never saw him fight at all (although, of course, he may have done so when we were not there). Mike maintained his position as alpha male for the next six years. Initially he was rather aggressive and worked hard to maintain his newly acquired status, but gradually he relaxed. The others soon accepted his authority without question, and he became an extremely benign boss.

As we look back over the thirty years at Gombe it becomes clear that, just as certain events mark the reign of a particular king in our own history, so it is with the reign of each alpha male. During Mike's reign we lived through the grim, bleak months of the polio epidemic. This epidemic began in the native population in the surrounding area. Two young men, diagnosed with poliomyelitis, lived in a small village near the southern boundary of the park. We assume the disease began there, but any visitor could have been a carrier. The first chimpanzee victim was Olly's four week old infant. He lost the use of all his limbs and seemed to be in pain, for whenever Olly started to move he cried loudly. Then she typically sat and cradled him, taking great care not to crush his limp arms and legs. I followed Olly for several hours. Often she stopped to cradle the crying infant. Presently she climbed a tree and stayed there for half an hour, grooming her six year old daughter, Gilka. When she climbed down the infant no longer cried out in pain and it seemed that he was dead (at the least he was completely unconscious). She continued to carry his body but how very different was her behaviour. Now, instead of tender nurturance, she treated the corpse as a 'thing', slinging it over her shoulders, dragging it by one leg, dumping it unceremoniously on the ground. I have seldom been more impressed than I was that day by the abrupt change in Olly's behaviour.

Another victim was Flo's eldest (or eldest known) son, Faben. He lost the use of one arm, but learnt to walk long distances in an upright position. And there were other victims: six, including Faben, were partially paralysed and learned to adapt to their disabilities; four were afflicted so badly that they died; two other probable victims disappeared at that time.

Mike was finally overthrown by Humphrey. By the time he lost power, Mike was looking old: probably he was somewhere around thirty five years old. His teeth, once so sharp and white, were worn and broken; his hair, once glossy black, was sparse and brown and brittle looking; and he seemed to have shrunk – his weight dropped from some eighty-five pounds to only seventy-eight. One day Humphrey, finding Mike alone in camp, charged and attacked him. Mike fled, screaming, Humphrey attacked him again for good measure, and the take-over was sealed. Mike dropped in rank very quickly after that, and ended up in a position low in the male hierarchy. It was no great glory, Humphrey's victory. He weighed at least twenty pounds more than Mike, and he could have defeated the ageing alpha long before. But he had continued to show submission, simply through custom, or habit. And even though he eventually dared challenge Mike and, with that one attack, became dominant over him, it was only through a lucky chance that he became the alpha of his community. For it was at that time, in 1970, that the community I had been studying for ten years began to divide. Leading the break-away group were two adult males of whom Humphrey was very much afraid. If they had not left, Humphrey would never have become alpha at all. As it was, he remained at the top for only eighteen months, for despite his large size and aggressive nature, he did not have what it takes to become a powerful alpha. He was really nothing but a big bully, with far more brawn than brain.

Humphrey was overthrown by Flo's second son, Figan, who weighed much less than his larger rival. But from adolescence on, Figan, like Mike, had been very highly motivated to improve his social status. And, also like Mike, he was intelligent. His tactics were different from those of Mike – all the cans had been stashed away. Figan, instead, made use of his close relationship with his older brother, Faben. The two travelled around together, and Figan spent much time grooming Faben. And he almost never challenged Humphrey unless Faben was in the same group – and then Faben almost always joined him. The two brothers then charged towards Hum-

David Greybeard, trusting enough to take a banana, page five.

Chimpanzees pant-hooting – their distinctive form of communication, page six.

Melissa greets Hugo with a kiss. Chimpanzees share greeting behaviours very similar to humans, page seven.

Social harmony is restored after a fight when the aggressor reassures the subordinated chimp with a touch, page eight.

A threatening swagger will usually eliminate the need for actual fighting, page eight.

Figan, who became alpha male with the help of his brother and remained dominant for ten years, page twelve.

Mike, who succeeded Goliath as alpha male through unusual intelligence and drive, page nine.

Three males patrolling the community's territory, page fourteen.

Males frightened while on patrol, page fifteen.

Chimps staring over alien territory, page fifteen.

Adult males can be very gentle and are usually paternal to all infants in their own community, page sixteen.

Weaning can be traumatic. Freud became clinging and depressed when Fifi weaned him, page nineteen.

Frodo is fascinated by sister Fanni, page twenty.

Freud carries baby Frodo, page nineteen.

phrey side by side, Faben with the magnificent upright display he had perfected since being stricken with polio. Over a period of about nine months Humphrey gradually became increasingly tense and nervous when in the presence of the two brothers, and Figan gradually became more secure, more confident. The show-down came one evening when Figan and Humphrey, as well as Faben, were in a large group. Suddenly, when many of the chimps were already in their sleeping nests, Figan began to display wildly through the branches. He leapt back and forth for some ten minutes, then plucked up the courage to drop down on Humphrey, who was already in bed. Humphrey, screaming, fell to the ground with Figan close behind. They grappled briefly, then Humphrey pulled away and ran off, still screaming. From that evening onwards Humphrey was submissive to Figan.

Figan, at least, for the first part of his reign, was the most powerful alpha male I have known during the thirty years. This was because he and Faben continued to spend most of their time together, each supporting the other. He had a rough time when Faben disappeared (we presume he died), because then the other males ganged up on him. But he always turned and faced up to their challenge; and they, for some reason, never pressed home their advantage, nor even when, as was sometimes the case, they were four to one. And so, for ten whole years Figan remained on top.

During this long reign, several events stand out. The two

that are most noteworthy are both somewhat grim, illustrating the 'dark' side of chimpanzee nature. Firstly, there was what we refer to as the 'four year war', secondly – and at about the same time – a four year period when two females, Passion and Pom, began to kill and eat newborn infants of their own community. (This second series of events will be described later.)

The 'four year war' occurred after the division of the community, referred to above. For a period of two years there were still occasional peaceful interactions between the males of the two sub-groups, but eventually we recognised two quite separate communities. The smaller Kahama community (seven males, three females and their young) took over a portion of land in the southern part of the territory that, previously, all the chimpanzees had shared. It was during the early part of Figan's reign that we observed the first in a series of raids that were perpetrated by the males of the northern or Kasakela community (nine males, thirteen females and their families) on the individual males and females of the Kahama chimpanzees. Those raids drove deeply into the heart of the territory which the break-away community had claimed as its own.

We know, today, that chimpanzees can be aggressively territorial. Each week, at Gombe, the adult males, in groups of three or more, tend to move out to peripheral parts of their home range, or territory (which at Gombe is usually about twelve to thirteen square kilometres), on what we call *patrols*. They start off with a fast and determined walk. Then, as they move into the overlap zone, between their community range and that of a neighbouring social group, they move very silently and cautiously, and much more slowly. They startle if they hear a sudden rustling in the undergrowth, and may reach out to touch or embrace each other, sometimes with big grins of fear on their faces. They may climb into a tall tree and sit staring out over what we can consider the hostile territory of a neighbouring community. They keep very close together, and so careful are they to avoid making a noise, that when walking through dry grass they may actually step from stone to stone, thus avoiding unnecessary rustling. If, during one of these patrols, they spy one or two individuals of the neighbouring community, they are likely to give chase. If they catch a victim they typically attack – and attack very severely. Attacks on community members, as mentioned above, are typically brief and seldom lead to injury. Fights on members of other communities are very different. They are brutal in the extreme, often involving several (up to six has been seen) males against one victim. These gang attacks usually last for at least ten minutes, and they typically result in very severe wounding. During 'the four year war' five savage assaults were actually observed; and we should remember that this brutality was directed against individuals who had previously been part of the same community. The victims of these attacks – which lasted up to twenty minutes – all died as a result of the terrible injuries they received. By the end of the four years the entire Kahama Community had been annihilated. Then the victorious northern males moved, with their females and young, into the part of their range that had been denied them when the Kahama chimpanzees had taken it over.

During our thirty years of research we have recorded more than twenty occasions when patrolling males attacked females of neighbouring social groups. Not adolescent females (who may actually be recruited when encountered) but mothers of small infants. When such a female is spied the males typically chase her and, if they catch her, attack. Sometimes they pound and hit and drag her one after the other, sometimes in unison. On four occasions these attacks have led to the deaths of the victims' infants, and three times the bodies of these infants were partially eaten. Moreover, when it was possible to get a good look at the females themselves, the nature of their injuries suggested that they would probably die also.

Whilst adult males are thus capable of the utmost brutality towards 'strangers', even including infants, they are typically gentle with infants of their own community. This only serves to drive home the fact that their aggressive intercommunity behaviour serves the purpose of protecting their territory and its resources for their own females and young. Indeed, the males of a community will co-operate to enlarge their territory at the expense of a weaker neighbour. The only chimpanzees, at Gombe, who can move freely from one community to another are adolescent females, or young adults who have not yet given birth. Indeed, patrolling males, as mentioned, may try actively to recruit such individuals if they encounter them near their boundaries. And these young females are sometimes only too willing to move into the territory of neighbouring males. Some emigrate permanently into the new community, others return, pregnant, and subsequently remain in their natal group. This is the main mechanism for avoiding too much inbreeding.

Within chimpanzee society it is often impossible to know which male fathered which infant. Sometimes, though, one male goes off with a female on a consortship. If he succeeds in keeping her away from other males throughout the last few days of her sexual swelling (at which time conception occurs), and if, some eight months later, she gives birth, then one can be assured that it was he who sired her infant. Many times, however, a female will be surrounded by most or all of the males of her community at the time of conception and the father of her infant is not then known. By and large, all males act in a paternal manner towards all the infants of their own community.

Chimpanzee infants, during the first three or four years of life, have a carefree existence. Not only their own mothers, but most familiar members of their group, show almost unlimited tolerance towards them. As we might imagine, there is considerable difference between females in the way they raise their families. There are 'good' mothers and 'bad' mothers, as is the case with humans. One female, whose life history we have recorded since she was an infant during the early days of the study, is Fifi. Her old mother Flo was a regular camp visitor from the very beginning, often arriving with David Greybeard and Goliath. Flo herself was a wonderful mother, tolerant, patient, affectionate and playful. It was not surprising to find that Fifi, when she had her first infant, showed many of the maternal characteristics that we had observed in Flo. When Fifi was an adolescent I did not realise that it is quite typical for young females to be fascinated by the sexual act, and I was much impressed by her obvious delight in the new experience: it seemed that she couldn't get enough! And so, when, at about thirteen years of age, she gave birth to her first infant, quite naturally I called him Freud.

From the start Fifi showed perfect cradling behaviour, supporting her infant securely so that he could relax and sleep without fear of falling. A less careful mother tends to provide poor support, and her infant is continually slipping off her lap and whimpering as he (or she) tries to get comfortable. In the wild, the chimpanzee infant nurses for five or even six years. For the first three years he suckles for about three minutes every hour, but thereafter suckling bouts are fewer and less regular.

As the infant gets older and begins to explore the world around him, he is always close to his mother, and she is ready to reach out and rescue him if he gets into difficulties. Young chimps often find it easier to climb up into a tree than they do to get down again. Like small human children they often get 'stuck'. Then they reach out, with small whimpers, asking for help from Mum – and the good mother is there to retrieve and rescue and comfort them. There is no time to follow Freud's early childhood in detail. When he was four years old Fifi began to make determined efforts to wean him: to prevent him from suckling and from riding always on her back during travel. For Freud, as for most infants, this period in his life was traumatic. He went into what we call weaning depression: the frequency of his play dropped, he spent much time in close contact with his mother, and sometimes he threw tantrums when she rejected him. However, by the time his young brother Frodo was born, Freud, by then five years old, was successfully weaned. The older child, however, continues to travel with his mother, and remains very emotionally dependent on her, for several years after the birth of a new infant. This means that he will develop close bonds with the new member of the family and just as Fifi had been fascinated by her young brother when she was a child, so was Freud fascinated by Frodo. As soon as their mother permitted it, Freud not only played with and groomed his young brother, but also carried him during family travel.

We learnt a great deal as we recorded Frodo's development and his interactions with the members of his family. It was particularly striking to note the tremendous difference between Frodo's early experience, and that of his elder brother Freud. For although Fifi had frequently responded to Freud's demands for attention, and frequently played with him, there had

nevertheless been many times when she was otherwise occu-
pied when her child had had to amuse himself. Moreover,
although Fifi had always been a social female, spending a good
deal of time with other adults, there were still long hours when
she, like all females, travelled and fed with only her infant for
company. For Frodo, life was very different. If Fifi was not
in the mood to play, Freud almost always was – a built-in
playmate, as it were. And it was easier for Fifi, too: when
Freud was small he was almost constantly pestering his mother,
for play or for grooming, whereas much of Frodo's need for
attention was fulfilled by Freud. Thus Fifi could sit peacefully
while her youngsters romped nearby; from her perspective,
Freud acted as a built-in baby sitter. Finally, Freud also served
as a built-in role model for his young brother. From the earliest
age Frodo watched the things Freud did and then often tried to
do the same himself. Thus he developed somewhat precociously.
When Freud was ten years old he quite often left Fifi for a
few days at a time to travel with the adult males, learning about
hunting and patrolling and courtship. And sometimes, for a
day or two, Frodo accompanied him on these excursions, even
though he was only five years old. Usually the young male does
not make his first such journeys until he is at least eight years
old. Frodo dared because he could travel with his brother, a
familiar family member to whom he could turn for comfort, for
reassurance, if things got tough. Even so, Frodo was still
spending almost all his time with Fifi when his sister Fanni was
born. And just as Freud had been fascinated by him, when he
was small, so was Frodo fascinated by Fanni. And Freud, who

still travelled very often with his family, developed close bonds with his infant sister also.

Five years later, Fifi, like clockwork, delivered her fourth child, her second daughter Flossi. It was not surprising to find that Fanni was utterly fascinated by her small sister. When Flossi was four years old, Fifi once more became pregnant. By this time Freud was eighteen and Frodo twelve years old. Both were spending far less time with Fifi, although the affiliative, supportive bonds between all family members were strong. Fanni and Flossi, though, still travelled constantly with their mother. Fifi started to wean Flossi almost a year earlier than her older children, and Flossi became very depressed. Nevertheless she was fully weaned when her infant brother, Faustino, was born, even though she was only four and a half years old.

Fifi is the most reproductively successful female we have known during thirty years at Gombe. From the age of about thirteen years she has produced a healthy infant within five years of each other and all her offspring have developed into large, healthy and well adjusted youngsters. Freud, the eldest is now twenty years old, and ten year old daughter Fanni could produce her firstborn within the next year or two. Friendly bonds between all family members remain strong.

For most females at Gombe, things do not go so smoothly. Indeed, over the thirty years of the study we find that it has been rare for a female to raise more than two offspring to full reproductive maturity – that is, until they are eleven to thirteen years old. For one thing there is high infant mortality at Gombe: about 30 per cent during the first year. One female whose reproductive history was very different from that of Fifi was Olly's daughter, Gilka. She was afflicted during the polio epidemic, losing partial use of one wrist. This meant that, when she began leaving her mother, she had difficulty in keeping up with groups of fast-moving adults, and after the death of her mother she spent much time alone. Then, during her adolescence, she developed a grotesque swelling of her face. It got so bad that we risked anaesthetising her so that a biopsy could be performed. Her condition turned out to be a fungus disease, quite common in West Africa but not previously described for East Africa. To some extent we were able to control this with medication. She was estimated to be twelve years old when she delivered her first infant: we were all very upset when he disappeared during his fourth week of life. About a year later she gave birth to a female baby. One day, as Gilka sat peacefully cradling her three week old infant, an adult female, Passion closely followed by her daughter Pom, suddenly charged towards Gilka and seized the baby. Passion bit deeply into the forehead, and then, carrying the dead body, climbed into a tree and began to eat the flesh. Her family, Pom and son Prof, followed and shared the grisely feast with their mother. Gilka, of course, had tried to protect her infant, but crippled as she was there was little she could do against these two large, aggressive and powerful females. Passion was top ranking in the female hierarchy, and Pom, as a consequence of her mother's rank, held a high position also.

That was the first observed instance of the cannibalistic behaviour shown by Passion and Pom that I mentioned earlier, and that occurred during Figan's reign as alpha male. With hindsight, I suspect that Gilka's first infant may have met the same fate. About a year after the observed killing, Gilka gave birth to a third infant – and Passion and Pom killed that baby also. Gilka fought even harder on that occasion, and received numerous severe wounds on her hands. These never really cleared up, and Gilka had increasingly difficulty in walking. She never became pregnant again, and died two years after losing her last infant.

We still have absolutely no idea as to why Passion and Pom behaved as they did. Had we known more of Passion's early life we might have gained some insight, but she was grown when I first knew her. She was always an asocial female, and had been a very harsh mother to her own first infant, Pom. It was only as Pom grew older that the very close bond developed between mother and daughter, and it was only because the two acted with such perfect co-operation that they were able to overcome some of the other females of their community. As it was, for a four year period those two females were observed to seize, kill and eat four new born infants. They never tried to attack a female if there were any adult males nearby, or even if another female was around – only when their prospective victim was by herself with her newborn. During the years of their rampaging, a total of ten infants died or disappeared; we suspect that most or all of them fell victim to Passion and Pom. Of the females in our study group, only Fifi managed to evade the killers and raise an infant.

Then Passion gave birth herself. Hoping that she would then be unable to continue her bizarre attacks we called the infant Pax. Subsequently Pom also gave birth, and with both females encumbered by small infants of their own, the extraordinary cannibalistic infant killing came to an end. Four years after giving birth to the aptly named Pax, Passion became extremely sick and was clearly in a great deal of pain. She continued to try to care for Pax, but as she became increasingly emaciated, it was obvious that most of her milk dried up. During this last illness she was accompanied by all her offspring: Pom (whose own infant had died), ten year old Prof, and infant Pax – a very close-knit family. When Passion died we did not expect that four year old Pax would survive. Previously, some infants of more than four years had seemed unable to recover from the psychological trauma of losing their mothers and had died shortly thereafter.

For a few weeks Pax was, indeed, very depressed. But, as he travelled with his two elder siblings, he gradually recovered. He formed a particularly close bond, initially, with his sister, Pom. An observer who did not know the true state of affairs might well have thought that these two were a mother-infant pair, except that Pom did not suckle Pax. After about a year, however, Pax transferred his allegiance to eleven year old brother Prof, and the two became all but inseparable. If they accidentally lost each other, Pax became extremely upset and continued to whimper, on and off, even if it was more than twelve hours before they found one another. And Prof, too, was clearly concerned if he suddenly realised that his small brother was not nearby. He would then leave any group he happened to be with and search in all directions until he and Pax were reunited. Today, Pax is ten years old, Prof sixteen, yet it is very rare to see either of them without the other.

Over the years, inevitably, we have recorded the behaviour of several infants following the loss of their mothers. Those with older siblings of either sex were typically adopted by them. One of the most remarkable series of events took place after the death of Miff, a female I had known since the early 1960s. Her son, Mel, was only just over three years old when she died during an epidemic, probably pneumonia, that claimed seven other chimpanzees of our group. Mel was a sickly infant. He had no elder brother or sister to care for him: two of Miff's other offspring had died, and her eldest daughter had emigrated into another community. It seemed impossible to us that Mel could survive. For three weeks he followed first one adult, then another. All were tolerant of the infant, but he had no special bond with any of them. And then, to our absolute amazement, he was 'adopted' by Spindle, a twelve year old adolescent male. Spindle waited for the infant during travel, allowed him to ride on his back and even cling to his belly if Mel was frightened, or if it was raining. He shared food in response to Mel's begging, and he allowed him to share his nest at night. He even ran to retrieve his small charge if Mel inadvertently got too close to a socially roused adult male, even though, on more than one occasion, this meant that Spindle got briefly hit or stamped on himself. There is absolutely no question but that Spindle saved Mel's life. Why did he act thus? Why did he encumber himself with a sickly infant with whom, previously, he had had almost no contact of any sort. Certainly they were not closely related. Probably we shall never know; there are many unresolved mysteries in our recorded life histories of the Gombe chimpanzees. Neverthe-less, it is perhaps pertinent that Spindle lost his mother during the same epidemic that claimed the life of Mel's. At twelve years old the young male, although he spends a great deal of his time travelling with adult males and sexually attractive females, very often seeks out his mother's company, particularly when things go wrong: when he gets hurt in a fight, for example. Perhaps there was a void in Spindle's life when he lost his old mother, a void that he tried to fill through his close association with the small and dependent Mel.

After a year Spindle and Mel began to spend less time together. And eventually Mel attached himself to the large and sterile female Gigi. Gigi, who has never given birth, has always been attracted to infants. During the past twenty years or so she has formed a series of relationships with one infant after another, as and when their mothers permitted. She was the perfect caretaker for the four year old Mel. Today he is seven years old (though no larger than a normal five year old). He follows Gigi almost all the time, is sometimes allowed to share her food, and very often gets support from her if he is threatened by another youngster. And he has a companion, for Gigi is acting 'Aunt' to another orphan, DarBee, who lost her mother during the same epidemic, and who is almost exactly the same age as Mel. For the past three years these two have been with Gigi almost all the time, except when she was in oestrus, travelling with the adult males. Then they often stay by themselves, two babes in the wood. During my last visit to Gombe I came upon Gigi in a tall tree, feeding on clusters of beautiful yellow blossom. Up above, the sky was blue, and through the branches of the tree I could see the blue waters of Lake Tanganyika twinkling in the sun. There, with Gigi, were Mel and DarBee – and a third orphan, little Dharsi, who had lost his mother about eight months before. He, too, had attached himself to Auntie Gigi. So that this female, who has had no infants of her own, is now serving as foster mother to three.

At the time of the epidemic, when Mel and Spindle lost their mothers, Figan was no longer alpha male. His place had been usurped by Goblin. There is no time to go into the fascinating details of Goblin's rise to the top. He accomplished this through sheer persistence. Again and again he would perform his spectacular charging displays towards and around groups of peacefully grooming or resting senior males. They tried to ignore him for as long as possible, then, often, ganged up to chase Goblin away. But though they sometimes attacked and even wounded him, Goblin never gave up. As soon as he had recovered he began again to perform his irritating, disruptive displays. In the end it almost seemed that they gave in for the sake of a quiet life!

I first knew Goblin in 1964 when he was just a few hours old, still attached to the placenta. He was Melissa's first infant and she seemed, for a while, quite puzzled by the tiny creature who had so suddenly come into her life. At first she seemed almost dazed, but soon became a proficient mother. Goblin grew up without complications and like other sons, maintained close supportive, affectionate bonds with his mother. Even as an adult Goblin could be comforted by an embrace from his mother. It is impossible to over-emphasise the value of friendly physical contact in maintaining friendly relations between the members of a chimpanzee community.

To digress for a moment. Melissa is the only female at Gombe known to have given birth to twins. We named them Gyre and Gimble. Right from the start one of them, Gyre, was clearly weaker than his sibling. Melissa was obviously not able to produce enough milk for the adequate nourishment of two infants and it was obvious that Gimble, the stronger of the two, was getting more than his fair share. Even so, when they were ten months old, they were both physically under-developed,

Fifi, here with all her family, was reproductively the most successful of all the chimps studied, producing a new baby approximately once every five years, page twenty-one.

Passion, first ranking female, page twenty-three.

Gilka with her infant that was murdered and eaten by two female chimps, page twenty-two.

Mel was one of a series of orphans taken over by the sterile female Gigi, page twenty-seven.

Goblin, successor to Figan as alpha male, rose to the top through sheer persistence, page twenty-seven.

Gyre and Gimble were the only twins known at Gombe. Only Gimble survived because mother Melissa's milk supply was insufficient for two, page twenty-eight.

Wilkie won a fierce fight with Goblin to become alpha male, suffering superficial wounds in the process, page twenty-nine.

Galahad, precocious and adventurous, may be destined to be alpha male one day, page thirty-one.

Clear cutting of trees poses the greatest threat to chimps as it destroys their habitat and heralds human encroachment, page thirty-two.

Olie and her four week baby, the first of twelve victims of a polio epidemic. The swelling on Ollie's neck is probably a goitre. page ten.

A baby chimp needs the same kind of love as a baby human does, but rarely receives it from captors, page thirty-three.

Cleo, though young and manageable, died from emotional neglect soon after this photograph was taken, page thirty-four.

Gregoire has lived in Brazzaville zoo since 1944, starving and hairless and performing his bizarre dance, page thirty-five.

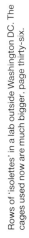

Rows of 'isolettes' in a lab outside Washington DC. The cages used now are much bigger, page thirty-six.

not much larger than normal two month old infants. That was when Gyre died of some respiratory disease. Gimble then began to develop rapidly, as though making up for lost time. But even today, at fourteen years old, he is still very tiny.

Goblin was sixteen years old when he became undisputed top ranking male of his community. He reigned for nine years, then was challenged by the number two male, Wilkie. They had a fierce fight which Wilkie, quite clearly, won. During the encounter both males were injured, but Wilkie's wounds, a gash in his upper lip and on one leg, were superficial. Goblin was less fortunate: he received many wounds on his hands and feet and, far worse, deep bite wounds in his scrotum which soon became infected. Goblin might well have died but, by chance, a British veterinarian, Kenneth Pack, arrived to help with the treatment of some sick baboons. He was able to immobilise Goblin, lance the wounds, and inject antibiotic into the blood. Subsequently we treated him four times a day with antibiotic powder inserted into bananas, and gradually he recovered.

During Goblin's long convalescence he was attended daily by one, sometimes two, of the Tanzanian field staff. They ensured that bananas with antibiotics were administered at the right time as well as noting all that happened. There was a time, prior to 1975, that students from around the world worked at Gombe. But then, as a result of a kidnapping incident, they had to leave. Since then, the day-to-day research has been taken over by a team of dedicated Tanzanian field staff: men from the villages around the national park. They learned to make detailed reports, and they now use 8mm video cameras. They are proud of their work, they discuss the chimpanzees with their families and friends, and they care about the chimps in the same way as I do. Sometimes they climbed high in the hills to get food for Goblin while he was sick: fruits that the other chimps were eating. I think that it is because these local people are so involved in the research that the Gombe chimps are safe from poaching.

I myself spent a good many hours with Goblin at that time. When I looked into his eyes I could not help but wonder: 'What is he thinking? Is he planning how he will get back into his group, face Wilkie and the other males?' Unfortunately for Goblin, he decided, when he was sufficiently recovered, to go charging back in his old style and try to pick up the reins of power. It didn't work. Not only did Wilkie attack him, but the whole group, including eight males and a female, joined in. No other observed intra-community aggression has been as fierce: it was like an assault on a stranger. Probably it happened because Goblin had been away so long: more than three weeks. Once again he received severe wounds, once again we started to administer antibiotics in bananas. And once again, eventually, he recovered. By this time he had learned his lesson: had accepted the fact that he was no longer alpha male. Initially he spent a good deal of time with the females, particularly his now adult sister Gremlin with her infant, Galahad, then finally worked his way back into the community. Today, although he is very submissive to Wilkie, the two are friendly and often travel together. And Goblin still spends much time with Gremlin and Galahad. And, I dare to predict, Galahad, precocious and adventurous, might well be alpha male himself in about fifteen years from now.

The Gombe chimpanzees, as I mentioned earlier, are in a very small national park, an area of only thirty square miles. Today this sanctuary is surrounded, absolutely, by cultivation and dwellings. Thus within the park the three communities of chimpanzees (not more than one hundred and sixty individuals) are isolated from any other chimpanzees remaining in the area. This is a pretty grim situation. In a population this size there is unlikely to be sufficient genetic diversity for long term survival. Nevertheless, the Gombe chimps, for the present, are at least safe from poaching. They live in a paradise compared to chimpanzees in most other parts of Africa. At the turn of the century there were chimpanzees living, in their tens of thousands, in the forested areas of West and Central Africa. Today the only really significant populations (numbering more than 5,000) are in Cameroon and Gabon, Zaire and Congo. Chimpanzees have vanished from four west African countries and have almost gone from five others. The remaining populations are dwindling fast as their habitats shrink. In the east the last remaining stronghold is Uganda: the other populations, in Rwanda, Burundi and western Tanzania, never huge, are fragmented and continually decreasing.

Chimpanzees are disappearing because of habitat destruction, for the great African forests are being destroyed just as they are everywhere else: trees are clear cut for growing crops and for human dwellings, and the timber trucks, mostly owned by greedy western or Japanese merchants, are driving ever deeper and deeper into once virgin forests. As roads are made, so, of course, the forest is opened up. Hunters and miners follow, then settlers. If you look at aerial photographs you will see how forests start being destroyed on either side of a new road. And people bring with them their diseases, another danger for the chimpanzees since they are susceptible to human contagious diseases. In many countries in Africa chimpanzees are hunted for food. And even in countries where they are not eaten, females are often selectively killed so that their infants can be captured and sold: sold as pets, or sold to dealers for the international entertainment and biomedical research industries. This method of killing mothers to get babies is not only cruel, it is also incredibly wasteful. Those who know most about the situation, estimate that for every infant that arrives at its final destination alive, *at least* ten chimpanzees will have died. Mothers creep away and die from their wounds; their infants will die also. Infants die of wounds they receive when their mothers are shot. Other infants die on the long journey from the forest to where they are destined to be sold, for not only do they lack proper care, but typically they are transported in tiny baskets, usually with hands and feet tied together with rope or even wire. Some of the infants that survive such treatment end up in dealer camps. Even here there are seldom people who understand the nutritional needs of their prisoners. One infant, needing his mother's milk, is given a few bananas and a dish of water. Above all, he needs his mother's love, but no one at the camp has any understanding of the psychological needs of infant chimpanzees. These are, in fact, virtually identical with those of infant humans. It is not surprising that the death toll mounts.

Nor is it surprising that caring people who come across one of these pathetic orphans huddled in a native market, or dragged from house to house on a string, should take pity on it, and rescue it. Of course, every time money changes hands, the trade is perpetrated. The proper course of action is to persuade the government to ban the sale of chimpanzees and to confiscate infants sold in violation of the law. And this is happening in many countries today. But irrespective of the rights and wrongs of the situation, the fact remains that many infant chimpanzees end up in private homes. If they are bought by people who care, and who understand something of their needs, the situation is not too bad for a while. Young chimps often live with the children of the family, they learn to eat at table, they sleep in their own little beds, they become part of the household. They make friends with the family dog. They enjoy their lives, unnatural though they may be.

But chimpanzees are not suitable as pets. What happens when they get older? When, at about five to six years of age, they are no longer cute and easy to control, when they become increasingly agile and can climb the curtains, when they can find the keys to cupboards no matter how well they are hidden, and unlock doors, and drink alcohol, and become ever more destructive of home and garden? What will their owners do with them when, as they get older, these young chimps increasingly resent discipline, are increasingly likely to bite, become potentially dangerous? Whenever people take chimpanzees into their homes the outcome is the same. At some point the youngster, for the second time, is deprived of its freedom. I have met so many like the female JoJo who, for several years, lived in a tiny cage, separated by a few bars from another tiny cage containing five dogs. The dogs are always barking, it's very noisy and very smelly. Four year old Cleo was really young enough to remain with her human family, but they didn't have time for her, so she was tied up outside, on a chain, and left for long hours by herself. She did not live long after the picture was taken: she couldn't cope with the psychological trauma that she had been exposed to in her few short years of life. Whiskey lived for two years, tied with a two foot chain to a pipe in a disused latrine at the back of a garage. He was owned by a man in Burundi, who referred to his prisoner as 'my son'. When I first met Whiskey, I made a commitment to him – to get him off his chain.

Gregoire was put into the Brazaville zoo, into a bleak

concrete, iron-barred cage, in 1944. He has lived there, by himself, ever since. When I first met him, a year ago, he was all but hairless, and his skin was stretched tightly across his bones. He could not see out of one eye. He showed a number of bizarre behaviours. Sometimes an African child would approach, stare at Gregoire, and say 'Dance!'. Then he would perform a strange pirouette and end up with his feet up on the bars. Usually the child then gave him a banana. There are zoos in many of the capital cities of African countries. It is hardly surprising that the chimpanzees are malnourished in countries like Zaire and Congo where the majority of the human population can afford only one meal a day. Unfortunately for the chimps, oranges and other fruits are good food for humans, too.

Some chimpanzee infants, far more than I used to think, are smuggled out of west Africa – particularly Equatorial Guinea and Senegal – to Spain and the Canary Islands. There, in the coastal resort areas, the pathetic youngsters are dressed up in unsuitable clothes and dragged around in the heat during the day; and round the nightclubs, in the noise and the smoke, at night. Many of the infants are sick, all are heavily drugged. We used to think they were drugged with valium, but we now know that they are often on hard drugs: when they are confiscated, many go through horrifying withdrawal symptoms. At five to seven years of age these chimps are no longer useful for the photographer's trade. Some have their teeth knocked out so that they can be used a little longer, but eventually even they become a liability. Then they disappear. We do not know what happens to them, though there is a suspicion that at least some end up in Eastern European medical research labs or zoos.

The United States uses more chimps for medical research than any other country. Prior to the United States signing the Convention on Trade in Endangered Species (CITES) hundreds of chimps were imported: now this is no longer possible. Chimpanzees used in medical research in the United States are bred in the laboratory. But for the young chimpanzee snatched from his mother it makes little difference whether she was in a lab or in the forest. The conditions in many such places are horrendous. One of them, just outside Washington DC (SEMA, Inc), has confined very many infants, in pairs, in cages measuring twenty-two inches by twenty-two inches, and twenty-four inches high. I saw some of them with my own eyes: they could barely move – unacceptable for any sentient being, let alone individuals whose needs are almost identical with those of human infants. Those that I saw had already been confined in these cages for more than three months. They had come from a breeding centre, and had a further three months to go. After quarantine they would be separated and placed singly in slightly larger cages which, at least until very recently, were put into 'isolettes' – small metal boxes with small panels in front that let in the light, like microwave ovens. Their only contact with the outside world, except during the brief periods when they were fed, cleaned or pricked with needles, was the sound of air rushing through noisy vents. The chimpanzees may remain within these prisons for three years or so. What happens to them, psychologically? Much the same as you would expect to happen if a young human child were so treated. The size of the smallest SEMA cages did not comply with the American federal regulations. But all over the United States and Europe and Japan there are adult chimpanzees in cages that *do* comply with the law, yet measure only five feet by five feet, and seven foot high. Chimpanzees can weigh up to one hundred and fifty pounds, can live for up to fifty years, and differ from us genetically by only just over one per cent. The cages are bleak and sterile. There are bars on all sides, above and, in some cases, below.

Chimpanzees are used in medical research because they are so like us, because physiologically they share so many similarities with ourselves – similarities in the composition of blood, the immune responses and so on. They are used in the search for cures and vaccines for diseases such as AIDS and hepatitis that other animals, less like ourselves, cannot be infected with. There is no concensus among scientists working in these fields as to whether or not chimpanzees are really useful, and I have spoken with many in Europe and the United States. Some say yes, others, equally emphatically, say no. I am not qualified to judge. But, leaving that question aside, we can fairly and squarely tackle the ethical question as to whether we are justified in keeping these close relatives in conditions like those described. They are guilty of no crime. Indeed, they may even be helping to alleviate human suffering. Yet they are sometimes kept in conditions far worse than those that we impose, today, on the most hardened criminals. Here it should be noted that not all labs are so grim; in some the chimps have much larger cages, live in groups, and may have access to outdoor areas. Moreover, in these labs they are being used for the same kind of research, which shows that the experiments can be carried out in a more humane way.

The inhumane conditions in so many labs developed because it was fashionable, at one time, to believe that animals had no minds, that they were simply bundles of stimulii and responses. It is a sad reflection on western science that whereas researchers pointed to chimpanzees as wonderful models for studying human diseases just because of their physiological similarities to ourselves, so many scientists have been reluctant to admit to the equally striking similarities between them and us in behaviour and emotions and intellect.

During the past fifteen years or so we have learned a great deal about the upper reaches of the chimpanzee mind. We have learned this from work in the field, and also from a good deal of research that has been done with captive chimpanzees – particularly, I think, the language acquisition experiments. We now know that chimpanzees are capable of intellectual feats that once were thought to be unique to humans. We know that they can reason, that they can plan for the immediate future, that they can solve simple problems. And we know that they can understand and use abstract symbols. Chimpanzees can be taught 300 or more signs of the American Sign Language of the deaf, ASL. They can use these signs in novel combinations in novel situations. Chimpanzees can even invent signs if they do not know the sign for a particular object. One youngster who was brought up with four older signing chimps, and who was never taught a sign by a human, had nevertheless used fifty-six signs in the proper contexts by the time he was eight years old.

This new understanding of chimpanzees' intellectual ability, along with observed similarities in emotional response between chimpanzees and humans and some of the behavioural similarities that have been described: the long childhood, the importance of social learning, the close bonding between family members that can last throughout a life of fifty years, the co-operation that may occur, the complex social structure, the altruism – all this not only teaches us a great deal about the chimpanzees' place in nature but about our place in nature too. It helps to blur the once sharply drawn line between humans on the one hand and non-human animals on the other. This is a little humbling because, whilst there is no doubt that we are a unique kind of primate, we are not quite as different as we used to think. We are not standing in isolated splendour on the top of a pinnacle, separated by an unbridgeable chasm from the rest of the animal kingdom. I believe that an understanding of the chimpanzee helps us to bridge, intellectually, this supposed chasm. And when the intellectual journey is complete, the bridge crossed, we cannot help but develop a new respect not only for chimpanzees, but for all the other amazing non-human beings with whom we share this planet.

This poses some tough ethical questions. Once we have this new respect for other life forms, what are we going to do in the face of the observed cruelty that our species perpetrates on supposedly 'lower' life forms? Not simply cruelty towards chimpanzees, but towards the animals we raise for food, those used, or abused, for entertainment, and, only too often, towards our 'pets'. What are we, as individuals, going to do about these things?

Finally, what can be done about the fact that animals are vanishing in the wild, in part because their habitats are being destroyed by humans? This, of course, is a huge problem. No-one can hope to do more than their small part, no-one can tackle more than a small aspect of this global problem. Our Institute (The Jane Goodall Institute) is still very small, but we are doing our best. We are working with African governments to set aside some forested areas for national parks, we are setting up sanctuaries for some of the orphaned chimpanzees in Burundi, Congo and Zaire. This next week in Burundi, Whiskey will be released from his chain and join JoJo and Cheeta and others in what we call our 'Half Way House' – large cages with connecting doors where up to fifteen other orphans will become acquainted and re-learn chimpanzee behaviour while waiting for their sanctuary to be built. We employed a keeper at the Brazzaville zoo to care for the primates: Gregoire has grown hair and put on weight and even his eye seems well again. We are even trying to do something for the chimpanzee prisoners in the medical labs. For me, my visits to the labs are like journeys into hell; but to see what goes on with my own eyes, to talk with the technicians, to try to raise awareness on the 'inside' as it were, they are journeys I must take. Ultimately we hope that it will be possible to introduce new legislation so that it will no longer be permissible, by law, to keep our closest relatives in tiny cages, alone.

Fortunately, all round the world, there are some truly wonderful people – there are some here tonight – who are working vigorously towards the same goals. Wherever there is a great need for help, such as in some far-flung African city, someone appears who is willing, and able, to do what is necessary. And there is a new awareness concerning the environmental and ethical problems that I have touched upon, particularly among students, not only in Europe and the United States, but in Africa too. This gives me great hope that we shall be able to make a difference.

Let me end with a story – for me a symbolic story. At the Detroit zoo in 1990 a new chimpanzee exhibit was opened, supposedly the biggest and best in North America. A water-filled moat surrounds the exhibit since chimpanzees cannot swim. Many chimpanzees were brought, from different parts of the world (so as to create a large gene pool), to join the two adult males already at the zoo. One of these, JoJo, had been there for some fifteen years, living most of the time in a small cage. Soon after the whole group had been released into their new enclosure, a fight broke out between JoJo and one of the new males. JoJo lost and ran into the water; he knew nothing of moats, of course. In his terror he managed to scramble over the barrier which had been built to prevent drownings. On the far side of the moat was a group of people, including one or two zoo staff. They knew that JoJo weighed 130 pounds, and that male chimpanzees can be dangerous, and so they stood and watched as JoJo sank for the third time. Luckily for JoJo there was a zoo visitor, a man who goes there once a year with his family. His name is Rick Swope. And *he* jumped in. He jumped in, in spite of everyone yelling and shouting at him. One of the visitors happened to have a video camera, and recorded the incident. It is one of the most dramatic tapes I have seen. The water is murky after rain, and Rick has to swim underwater searching for JoJo's limp body. Somehow he gets the 130 pound deadweight up onto his shoulders, and somehow he manages to get him over the safety barrier. And all the while the humans watching are shouting at Rick: 'Get out, get out!', 'Sir! that chimp will kill you', 'Let the monkey drown', 'You fool, get out!' But Rick pays no attention, and finally pushes JoJo onto the shore of the exhibit. The bank is steep (much too steep) and when Rick lets go, JoJo, who is still unconscious, slips back towards the water. As Rick pushes him up again, the yells from the spectators are redoubled, for now some of the other adult chimpanzees are rushing towards the scene, probably to 'rescue' JoJo. Rick looks back towards the shouting humans, glances at the bristling screaming chimps above him, and again lets go of JoJo, who again slides down the bank. 'Let the ape drown!', 'Sir! you can't stay there', 'The monkeys will kill you', 'Rick, come back, come back'. The calls are frenzied now. But Rick pushes JoJo back up the bank and stays there, supporting the inert body. And then comes the wonderful moment when JoJo raises his head, shakes it a little, and very groggily moves a few paces up the slope and collapses on more level ground. Rick saved JoJo's life.

Later, the director of JGI (USA) spoke to Rick: 'That was a very brave thing you did. You must have known it was dangerous. Why did you do it?' After a pause, Rick said 'Well you see, I looked into JoJo's eyes before he went down the third time. And it was like looking into the eyes of a man. And the message was: "Won't anybody help me?"'

For me this story symbolises all those occasions when we find *ourselves* standing on the brink of muddy water and are afraid to jump in. Next time we can ask ourselves: are we going to stand here, like the zoo staff, and watch, because we think it is dangerous, or are we going to be like Rick, and jump in, and do our best to make a difference?

Thank you.

APPENDIX I

INTRODUCTION BY ELEANOR McLAUGHLIN, LORD PROVOST

Citation

The Edinburgh Medal has been instituted by the City of Edinburgh District Council to honour men and women of Science who have made a significant contribution to the understanding and well-being of humanity. It is awarded in 1991 to Jane Goodall for her initiation and development of one of the most important scientific projects of this century. In this work she has demonstrated a model of research that is able to grasp the richness of life and, in her descriptions of chimpanzees, she has held up a mirror to humanity: both we and they are the beneficiaries.

———————————— ◇ ————————————

It is with great pleasure that I welcome here tonight, on behalf of the City of Edinburgh, Jane Goodall, to receive the Edinburgh Medal.

It is now over five years since the idea of the Edinburgh Medal and the International Science Festival was first proposed. This is now our third Festival and I have no doubt that it will have exceeded all expectations. When we finish on this coming Sunday over 200,000 people will have participated in one way or another.

Nor is it just in numbers that the Festival has grown. The scientific quality and depth of the Festival also improves, particularly I believe in what is the special quality of the Festival: its encouragement of critical questioning and debate.

This year has also seen for the first time the publication of a separate programme for schoolchildren with almost as many events as were included in our first Festival. This, if it didn't succeed in attracting children from New York, as we did last year, did bring in many from England and Scotland, from Wigtown to Fort William.

Last year I described our City's launching of the Festival as a 'step in the dark', it is more than clear now that it is becoming the international beacon we hoped it would.

This is shown not only by the rapidly expanding international participation but also by the visits of delegations from Belgium, Italy, New Zealand and Korea to investigate what we do and how we do it, with the intention of repeating our success.

This is praise indeed and to be soundly welcomed but I would be less than human if I were not also slightly concerned about our unique event being replicated elsewhere and gaining financial support from commerce and government that we are as yet unable to command. Because, if there is a blot on the horizon of the Science Festival, it is the weakness of commercial sponsorship received this year and which remains very weak in our planning for 1992 and beyond.

I hope I will be forgiven for that digression although I don't doubt that Doctor Goodall more than understands the difficulties of raising funds for critical enquiry and the development of understanding.

The Edinburgh Medal has been established by the City Council to honour scientific workers for their contribution to both science and society. Without doubt Jane Goodall's work has been one of the great scientific projects of this century. It was commenced with no official scientific training or qualifications, and we should not hide from ourselves that that may well have been an advantage. In its practice, her work has shown that the deepest understanding is not gained by reducing things into separate, isolated, even abstract components, but by the drawing together of all elements into an active pattern of change and development.

Such a reaffirmation of a rich and productive scientific method is to her own great credit. In addition, the product of that method has revolutionised our understanding of humanity's place in nature and even of what society is, for the results have shown that, in contrast to the common view of humans being the very apex of creation, there are other tool-using, communicating, social primates, the chimpanzees, right alongside us.

This knowledge has been conveyed to us not only in scientific papers and monographs but in books for both children and adults which, whilst rigorously scientific, are also readily understandable. These publications and her work through the Jane Goodall Foundation have been very influential in bringing about the improvement of conditions for primates in both captivity and the wild.

It seems to me important in a period when some have been glorifying the technology of mass murder that tonight, to borrow and amend Stephen Jay Gould's phrase, 'we celebrate the glory of the primate intellectual tradition' in one of its finest manifestations.

APPENDIX II

ORATION

Steven Rose

It is a rare privilege to be able to introduce to you tonight a biologist with a name to conjure with, one whose life and research have become almost mythologised and certainly made more widely familiar than that of any other living primatologist by virtue of film, television and popular writing. Although, like others, I have known of her work for many years, I meet her face to face as a fellow biologist for the first time tonight. We represent, I suspect, widely different regions of the research – and possibly political – spectrum. As to the former, unlike Jane Goodall, my work is confined to the laboratory, and to the dissection of the minute molecular processes that underlie behaviour. The understanding of this is, however, of passionate concern to both of us. As to the latter, we will doubtless discover more as she begins her theme in a few moments.

Dr Jane Goodall has at least four careers: as a meticulous observer of chimpanzee life in the wild; as a committed campaigner against the inappropriate and excessive uses of animals in research; as an author of popular books about both her chimpanzees and other animals; and as a living myth. Reality and myth converge on a life history which begins in 1934, with a conventional south of England, middle-class upbringing. This fitted her to become, as was appropriate for nice young girls of her class and period, a secretary. Going to Kenya on holiday in her early twenties, she met the anthropologist Louis Leakey, became his secretary and later researcher. Beginning in 1960 she started her pioneering work studying chimpanzees in the Gombe Stream Game Reserve on the shore of Lake Tanganyika. In the mid 1960s she achieved the almost unique distinction of becoming a Cambridge PhD without any first degree – a tribute to her scholarship, determination, and perhaps also to the congenial research ambience created by her supervisor, Robert Hinde.

From that time on, undeterred by the political and social upheavals which have characterised the African states around her chosen sites – upheavals resulting in violence and the kidnap and even murder of other animal researchers – she has continued to act as the loving social and personal biographer of her chimpanzee subjects. As anyone who has studied animals will know, such work is often immensely tedious, it requires teamwork and meticulous observation and, to convert observation into reputable scientific data, complex statistical procedures. To aid in this work, over the years Jane Goodall has built up a team of observers and assistants, both African and foreign. The members of this team now possess between them an impressive array of doctorate degrees and university professorships across the globe. This collective work, and the research papers that have emerged from it, have greatly extended scientific understanding of the individual and social life of humanity's closest extant relative. We now have a rich and complex picture of chimpanzees as being possessed of unique personalities, bonded into a web of psychological, social, economic and even political relationships.

It has been Jane Goodall's gift of turning detailed observational fieldwork into highly readable popular accounts, in such books as In the Shadow of Man, The Chimpanzees of Gombe *and* My Life with Chimpanzees *which have made her into more than just a brilliant observational primatologist. They have launched her on her additional career as science writer and provided the launchpad for her campaigning conservationism. They also began the transformation of the researcher into the living myth perpetuated by a series of films made under the auspices of* National Geographic. *It is a dangerous thing to achieve mythic status, for the process inevitably tends to obliterate the collective nature of research work and instead offer the somewhat misleading picture of a heroine-scientist, a lone white woman in the jungles of Africa.*

It is, however, not the myth but the scientist that Edinburgh is honouring by the award of its medal this evening. The only condition attached to this award is that the recipient uses the occasion to speak under the general heading of science and society, a theme dear to my heart, and it will not, I think, surprise, but it will delight, everyone here this evening to see that Dr Goodall has entitled her lecture: the chimpanzee – the living link between 'Man' and 'Beast'. *She has placed quotation marks round the two crucial terms of that title, for reasons which we all eagerly anticipate.*

APPENDIX III

VOTE OF THANKS

Aubrey Manning

I think that Jane Goodall's work has profound significance for us in a number of ways. I think the first way was for biologists as she really showed us that it was possible. Until news began to emerge of her results in Gombe, most of us would not have thought that it was feasible to gain such information from wild animals. As a result of her pioneering achievements an enormous number of studies have been carried out on many other primates, but she was the true leader.

To do such work required great physical courage, and I think that's been fairly obvious from the last story she told us. Her work also has profound significance in another way: it has shown us that chimpanzees are magnificently themselves. I don't think, frankly, that we have much to learn in behavioural terms from chimpanzees. What we can learn from them is the much more profound lesson that they are themselves and we ought to give them the space to be themselves. For me this is really the most important lesson of all.

As Jane has so eloquently told us, it will require a great deal of work and sacrifice from us if the chimpanzees are to survive, and survive in the way that they deserve. She has shown us what they require and, in doing so, has revealed some hints of how to restore that vital balance between humanity and the other living things with which we share the planet. The lesson is important, obviously, for the chimpanzees who are greatly and immediately threatened but I think the point is also one for our own survival.

In accepting the Edinburgh Medal and coming to speak to us tonight I think we all have gained a tremendous amount: we can congratulate not only Jane Goodall on a magnificent lecture but ourselves on our good taste and good judgement. She is a pioneer – her work is of such importance that it will always live and tonight we have been privileged.

APPENDIX IV

THE JANE GOODALL INSTITUTE (UK)

The Jane Goodall Institute (UK) is committed to a continuing effort to learn more about chimpanzees and the extent to which they are exploited, and to sharing this information with as wide a public as possible.

Its aims are:

- *to support and expand research into the behaviour of chimpanzees and other non-human primates;*

- *to conduct research throughout the rapidly decreasing range of chimpanzees in Africa that will form the basis of advice relevant to the setting up and maintenance of areas for the protection of selected chimpanzee populations;*

- *to investigate the effects of the various physical and psychological deprivations inflicted on chimpanzees and to study the effects of environmental enrichment on the behaviour of such chimpanzees with a view to ascertaining and promulgating optimal conditions for the maintenance of chimpanzees in captivity;*

- *to construct and maintain facilities in order to monitor behaviour changes that occur when chimpanzees are removed from inappropriate captive conditions and placed in groups in suitable enclosures;*

- *to make known to as wide a public as possible, through various publications and through the media, the results of all projects undertaken.*

Further details and information on membership can be obtained from:

> *Robert Vass (Secretary),*
> *Jane Goodall Institute (UK),*
> *15 Clarendon Park,*
> *Lymington,*
> *Hants SO41 8AX*

Published by
EDINBURGH UNIVERSITY
PRESS

on behalf of
THE CITY OF EDINBURGH
DISTRICT COUNCIL

in association with
THE EDINBURGH
INTERNATIONAL FESTIVAL OF
SCIENCE AND TECHNOLOGY

Design contribution by
TAYBURN DESIGN GROUP
EDINBURGH